Famous Flights

Understanding and Using Variables

Greg Roza

Math for the **REAL World**™

Rosen Classroom Books & Materials™
New York

Published in 2006 by The Rosen Publishing Group, Inc.
29 East 21st Street, New York, NY 10010

Book Design: Michael J. Flynn

Photo Credits: Cover, pp. 6, 11, 13, 17, 19, 21, 23, 25 (Jerrie Mock) © Bettmann/Corbis; p. 5 © Archivo Iconografico, S.A./Corbis; pp. 9, 15, 25 (Wiley Post) © Hulton Archive/Getty Images; p. 25 (Lucky Lady II) © Time Life Pictures/Getty Images; p. 27 © Jim Sugar/Corbis; p. 28 © Ruber Sprich/AFP/Getty Images.

Library of Congress Cataloging-in-Publication Data

Roza, Greg.
  Famous flights : understanding and using variables / Greg Roza.
      p. cm. — (Math for the real world)
  Includes index.
  ISBN 1-4042-3367-9 (lib. bdg.)
  ISBN 1-4042-6087-0 (pbk.)
  6-pack ISBN 1-4042-6088-9
  1. Algebra—Juvenile literature. 2. Variables (Mathematics)—Juvenile literature. 3. Air pilots—Juvenile literature. I. Title. II. Series.

QA155.15.R69 2006
512—dc22
                                                    2005017780

Manufactured in the United States of America

Contents

What Are Variables?

An algebraic equation includes known numbers, called constants, and unknown numbers, called variables. Variables are usually represented by letters. To set up or solve an algebraic equation, first determine what the variable or variables stand for. You also may need to rearrange the equation in order to find a variable's value.

Here's a simple example. Let's say that you made a paper airplane and took it to the park to see how well it could fly. You threw the plane 3 times. During the first flight, the plane flew 15 feet. During the second flight, the plane flew 13 feet. The total distance of all 3 flights was 47 feet. How can we find the distance of the third flight? Follow the steps in creating and solving the equation on page 5.

Luckily for us, people throughout history have found ways to make objects other than paper fly much farther distances. Now let's use what we know about variables to learn about some of these famous flights.

Algebra was invented around A.D. 830 by an Arab mathematician named al-Khwarizmi. He invented algebra in part to help merchants solve math problems they encountered every day in marketplaces such as this.

First, list the information we know:

first flight = **15 feet**

second flight = **13 feet**

total distance = **47 feet**

The distance of the third flight is unknown. Let's represent it with the variable x.

first flight + second flight + third flight = total distance

15 + 13 + x = 47

First, add 15 and 13.

28 + x = 47

Then subtract 28 from both sides of the equation to keep them equal.

28 – 28 + x = 47 – 28

This will get the variable alone on 1 side of the equation and reveal the answer to the question.

x = 19 feet

The third flight was a distance of 19 feet.

The First Flight

For centuries, people have tried to create flying machines. During the late 1700s, inventors began experimenting with hot-air balloons. Hot-air balloons are capable of flight because the heated air in them is lighter than the air surrounding them. Although these balloons allowed people to fly, they were difficult to control and nearly impossible to steer.

On June 4, 1783, 2 wealthy French brothers—Joseph-Michel and Jacques-Étienne Montgolfier—

demonstrated that filling a balloon with hot air causes the balloon to rise. In September 1783, the Montgolfier brothers placed a sheep, a duck, and a rooster in a basket suspended beneath a hot-air balloon and sent it into the air. The balloon floated for about 8 minutes, rose to about 1,800 feet (549 m), and landed safely 2 miles (3.2 km) away. This was the first successful flight involving living creatures.

The first human hot-air balloon flight occurred on November 21, 1783. Jean-François Pilâtre de Rozier and François Laurent flew in a balloon designed by the Montgolfiers. It traveled a distance of about 29,040 feet (8,851 m) over Paris, France, at a speed of about 1,161.6 feet (354.1 m) per minute. How long did this historic flight last?

We know 2 pieces of information:
distance (29,040 feet) and speed (1,161.6 feet/minute).

Let's use the formula for distance to find out how long the flight lasted.
distance = speed x time
$d = s \times t$

Place the information we know into the equation.
29,040 feet = 1,161.6 feet/minute x t

Now we need to solve for t by dividing both sides by 1,161.6. This will get the variable alone on 1 side of the equation.

$$\frac{29,040}{1,161.6} = \frac{1,161.6 \times t}{1,161.6}$$

$$\frac{29,040}{1,161.6} = t$$

t **= 25 minutes**

Before dividing 2 numbers, the divisor must be a whole number. Change the divisor into a whole number by moving the decimal point to the right. Be sure to move the dividend's decimal point the same number of places to the right.

29,040 ÷ 1,161.6 ⟶ 290,400 ÷ 11,616

```
                25
    11,616 ) 290,400
            −232 32
              58 080
            −58 080
                   0
```

The first manned hot-air balloon flight lasted about 25 minutes.

The Wright Brothers

Prior to the 1900s, many inventors attempted to build airplanes, but all failed. An airplane is a heavier-than-air machine that uses a power source and wings to stay in the air. It also needs a system to control the plane's movements.

Many people think of Orville and Wilbur Wright when they think of the first airplane. However, they were not the first people to construct a plane. They were the first to solve problems that had prevented earlier airplanes from making successful flights. Earlier airplanes could not stay in the air or could not be controlled while in the air. The Wright brothers discovered that an airplane must be capable of 3 types of movement to allow a pilot to control it: roll, pitch, and yaw. A plane rolls when 1 wing is higher than the other. Pitch moves a plane's nose up or down. Yaw moves a plane's nose left or right.

A simple plane—such as the plane the Wright brothers built—is capable of rolling, pitching, and yawing due to a series of levers, cables, and pulleys. The levers control several basic parts of the plane, allowing it to move up, down, left, and right. The rudder on the tail of the plane controls yaw. The tail also has flaps called elevators that control pitch. The Wrights warped the shape of their plane's wings to control rolling. Later designers developed flaps on the wings called ailerons, which allow the same control. The throttle—which is part of the engine—handles speed and power. Before the Wright brothers had discovered these airplane basics, successful flight was not possible.

roll pitch yaw

The Wright brothers were successful in many areas. Before they turned 30 years old, they had made and sold small toys, made and sold bicycles, and built a printing press and used it to start their own newspaper.

Perhaps the most famous flight in history was the Wright brothers' first airplane flight. That event proved to the world that heavier-than-air flight was truly possible. It also inspired other inventors to improve upon the first successful airplane.

On December 17, 1903, the Wright brothers took their plane—the *Flyer*—to an empty field near Kitty Hawk, North Carolina. Orville was first to take the controls. This flight lasted only 12 seconds. The plane traveled 120 feet (37 m) before landing smoothly. After centuries of failure, man had finally succeeded in creating a heavier-than-air flying machine that could be controlled while in flight.

The Wright brothers made a total of 4 flights that day. Wilbur was pilot for the fourth and longest flight, which lasted 59 seconds and covered 852 feet (260 m). The Wright brothers flew a total of 1,347 feet (411 m) in all. Using the information presented here, we can find the average distance of the second and third flights.

Orville and Wilbur had only 5 witnesses that day: 4 men and a boy. One of the men took the picture shown here just as the *Flyer* left the ground for the first time.

Here's our information:

first flight = 120 feet
fourth flight = 852 feet

total distance = 1,347 feet
second and third flights = m

First, set up an algebraic equation for the total distance the *Flyer* traveled.

first flight + second and third flights + fourth flight = total distance
120 feet + m + 852 feet = 1,347 feet

972 + m = 1,347
972 − 972 + m = 1,347 − 972
m = 375

Add the first flight and the fourth flight together. Then subtract that number from both sides of the equation. This will get the variable alone on 1 side of the equation.

375 ÷ 2 = 187.5 feet

Added together, the second and third flights equal 375 feet. To find the average of these 2 flights, divide 375 by 2.

The average of the second and third flights is 187.5 feet. The actual distances for these 2 flights were 175 feet and 200 feet.

Crossing the Atlantic

In the years after the Wright brothers made their historic flight, others began to see the importance of more powerful airplanes that could fly farther and longer.

Toward the end of World War I, the U.S. Navy wanted an airplane capable of crossing the Atlantic Ocean. The Navy and the Curtiss aircraft company built 4 "flying boats" that were able to land on water. The NC (Navy-Curtiss) aircraft were called the NC-1, NC-2, NC-3, and NC-4. The NC-2 was damaged, and its parts were used as spare parts for the other planes. The other 3 planes began the NC **Transatlantic** Flight Expedition on May 8, 1919, taking off from the Rockaway naval station on Long Island, New York. The planes crossed the Atlantic Ocean in several "hops," stopping to refuel. NC-1 and NC-3 both became damaged after landing in ocean waters to avoid flying through bad weather. Only 1 of the planes, the NC-4, made it to Lisbon, Portugal, on May 27, 1919. The NC-4 continued on to Plymouth, England, on May 31. After 3,322 miles (5,345 km), the NC-4 and its crew had completed the first transatlantic crossing.

The first transatlantic flight took 19 days, but only about 5.8% of that time was spent in the air. How many hours was the NC-4 in the air?

This picture shows the NC-4 just after it landed in Portugal's Lisbon Harbor on May 27, 1919. The NC-4 has been restored and can be seen today at the National Museum of Naval Aviation in Pensacola, Florida.

First, let's list the facts that we know:

total time of trip = 19 days, or 456 hours

approximate percent of time in flight = 5.8%

The variable will be the total number of hours in the air, a.

To find out how many hours the NC-4 was in the air, we can set up an algebraic equation. Since we are working with a percent, the equation can be a proportion. A proportion is a comparison of 2 ratios or fractions.

$$\frac{a}{456} = \frac{5.8}{100}$$

To solve for a, we need to cross multiply these fractions.

$$100 \times a = 5.8 \times 456$$
$$100 \times a = 2{,}644.8$$

$$\begin{array}{r} 456 \\ \times\ \ 5.8 \\ \hline 364\ 8 \\ +\ 2\ 280 \\ \hline 2{,}644.8 \end{array}$$

$$\frac{\cancel{100} \times a}{\cancel{100}} = \frac{2{,}644.8}{100}$$

Now we need to get the variable a alone on 1 side of the equation. To do this, divide both sides by 100. This will give us our answer.

$$a = 26.448 \text{ hours}$$

Altogether, the NC-4 was in the air for 26.448 hours, or for about $26\frac{1}{2}$ hours.

Charles A. Lindbergh

The success of the NC-4 caused a stir in the world of **aviation**. Airplane manufacturers worked to improve upon the NC-4's design. Other pilots attempted transatlantic flights. A few were successful, but others were not able to finish the journey. Some pilots even died in the attempt. In 1919, a New York City hotel owner named Raymond Orteig offered $25,000 (a value of about $280,000 today) to the first pilot who could fly solo nonstop from New York to Paris, France. Several pilots tried unsuccessfully to win the prize.

In 1927, a young pilot named Charles A. Lindbergh believed that he could win the prize money with the right plane. Lindbergh convinced 9 businessmen from St. Louis, Missouri, to fund the project. With Lindbergh's help, the Ryan **Aeronautical** Company of San Diego, California, designed and built a plane that Lindbergh named the *Spirit of St. Louis*. On May 10 and 11, 1927, Lindbergh tested the plane by flying it solo from San Diego to St. Louis, and then from St. Louis to New York City. This flight set a transcontinental record of 20 hours and 21 minutes. Lindbergh then prepared himself for a nonstop transatlantic flight.

In 1925, Charles Lindbergh began delivering mail by plane between St. Louis, Missouri, and Chicago, Illinois. He also had experience as a barnstormer, or a pilot who performed stunts at fairs.

At 7:52 a.m. on May 20, 1927, Lindbergh took off from Roosevelt Field near New York City. Loaded with 451 gallons (1,707 l) of fuel, the *Spirit of St. Louis* flew northeast and followed the coast all the way to Newfoundland, Canada. From there, Lindbergh steered his plane out over the Atlantic Ocean. He saw storm clouds ahead of him, and he veered slightly away from them when he noticed ice forming on the plane's wings. Much of the flight was made through heavy fog. Lindbergh had to struggle to remain awake. He even fell asleep several times during the long flight. At one point, he flew just 10 feet (3 m) above the ocean to keep himself awake.

The fog kept Lindbergh from seeing land. Eventually he knew that he was near the coast of Ireland because he spotted fishing boats. He even tried to ask the fishermen for directions, but they could not hear him. The weather cleared as he flew over Ireland and England. Then—after $33\frac{1}{2}$ hours and 3,610 miles (5,808 km)—Lindbergh landed safely at Le Bourget Field in Paris, France.

If the *Spirit of St. Louis* used about 10.9 gallons of fuel for each hour of the flight, how much fuel was left in its tanks by the end of the flight? Let's use algebra to find the answer.

Let's list the information we will use:

length of flight = 33.5 hours **total amount of fuel = 451 gallons**
amount of fuel per hour = 10.9 gallons **fuel remaining = f**

We can find out how much fuel was remaining by setting up a simple algebraic equation.
fuel used + fuel remaining = total amount of fuel

Although we don't know how much fuel was used, we can find out by multiplying
2 pieces of information we have:
fuel used = amount of fuel per hour x number of hours
(amount of fuel per hour x number of hours) + fuel remaining = total amount of fuel

(10.9 x 33.5) + f = 451 ←

$$\begin{array}{r} 10.9 \\ \times\ 33.5 \\ \hline 5\ 45 \\ 32\ 7 \\ +\ 327 \\ \hline 365.15 \end{array}$$

365.15 + f = 451
365.15 – 365.15 + f = 451 – 365.15
f = 85.85 gallons

The *Spirit of St. Louis* had about 85.85 gallons of fuel left after crossing the Atlantic Ocean.

Amelia Earhart

Amelia Earhart was the first famous woman pilot and one of the most memorable pilots of all time. Earhart had been a volunteer nurse during World War I in a hospital in Canada. After the war, she moved to California to be with her mother. There she became interested in aviation. Earhart took flying lessons with one of the few female pilots in the world, Neta Snook. Earhart purchased her first plane—a used, bright-yellow **biplane** that she named "Canary." On October 22, 1922, she used this plane to set a women's record by flying to an **altitude** of 14,000 feet (4,267 m).

In 1928, Earhart was invited by book publisher George P. Putnam to join a crew for a flight across the Atlantic Ocean. Earhart became the first woman to cross the Atlantic in an airplane on June 18, 1928. This made Amelia Earhart a worldwide celebrity and convinced her to dedicate all of her time to aviation. In 1929, Earhart helped to found a group called the Ninety-Nines, the first international organization for women pilots. Today, the Ninety-Nines still helps women all over the world learn to fly and to gain employment in the field of aviation.

Amelia Earhart married George P. Putnam, pictured here, in 1931. Putnam helped Earhart plan her famous solo transatlantic flight.

By 1932, Amelia Earhart knew it was time to tackle crossing the Atlantic alone. On May 20, 1932, at 7:12 p.m., Earhart took off from Harbour Grace, Newfoundland, in a **monoplane**. She had originally intended to land in Paris. However, she experienced difficulties during the flight that forced her to change her plans. At one point, her plane went into a spin and dropped nearly 3,000 feet (914 m) before she pulled out of the spin. As soon as Earhart was above Ireland, she began looking for a place to land. On May 21, about 15 hours and 18 minutes after take off, Earhart safely landed in a cow pasture near Londonderry, Northern Ireland.

Amelia Earhart had become the second person—and the first woman—to cross the Atlantic alone by airplane. For this accomplishment, she became the first woman to be awarded the **Distinguished** Flying Cross by the U.S. Congress. She received other honors as well.

Amelia Earhart continued to break aviation records. On January 11 and 12, 1935, Earhart was the first person to make a solo flight from Hawaii to California. This flight across the Pacific Ocean was 382 miles (615 km) longer than her flight across the Atlantic. However, the flight across the Pacific was 1,202 miles (1,934 km) less than Lindbergh's 3,610-mile (5,808-km) flight across the Atlantic. How long was Earhart's Atlantic flight?

In this photograph taken in 1932, Amelia Earhart receives the National Geographic Society Medal from President Herbert Hoover. Tragically, Earhart disappeared over the Pacific Ocean in 1937 while attempting to fly around the world.

First, list the information we will use:

Lindbergh's Atlantic flight: 3,610 miles
Earhart's Atlantic flight = a
Earhart's Pacific flight = 382 miles + a
Earhart's Pacific flight = 3,610 miles − 1,202 miles

Use the information about Earhart's Pacific flight to create an equation and solve for a, the distance of the Atlantic flight.

$3,610 − 1,202 = 382 + a$
$2,408 = 382 + a$ ← Solve this equation by getting a alone on 1 side.
$2,408 − 382 = 382 − 382 + a$
$2,026$ miles $= a$

Earhart's flight distance across the Atlantic was 2,026 miles.

Breaking the Sound Barrier

In the years following Earhart's famous flight, scientists began experimenting with new types of **technology** that allowed planes to fly farther, faster, and higher. The first successful jet-engine airplane was flown in Germany in 1939. Soon other countries were building and testing their own jet airplanes.

A young U.S. soldier named Charles "Chuck" Yeager earned a reputation as an excellent pilot during World War II. After the war, Yeager became a flight instructor and test pilot for the newly formed U.S. Air Force. In 1947, the Air Force asked Yeager to test a new jet plane—the X-1 fighter plane. This plane was designed to test the "sound barrier." At the time, many people believed that it was impossible to fly faster than the speed of sound, which is about 761 miles (1,224 km) per hour at sea level in temperatures around 60°F (16°C). The speed of sound changes with different altitudes and air temperatures.

On October 14, 1947, Yeager flew the X-1 at a speed of 700 miles (1,126 km) per hour at an altitude of 42,000 feet (12,802 m). He set a new world speed record by flying the X-1 about 1.06 times faster than the speed of sound. What was the approximate speed of sound for this altitude? Round your answer to the nearest mile per hour.

On December 12, 1953, Chuck Yeager (right) set another world record when he flew a jet plane at a speed 2.44 times faster than the speed of sound at the altitude he was flying. His top speed was 1,650 miles (2,655 km) per hour at about 76,000 feet (23,165 m).

Here's the information we know:

Yeager's top speed = 700 miles per hour
how many times the speed of sound = 1.06
speed of sound at 42,000 feet = s

To find the speed of sound at 42,000 feet, set up an algebraic equation. Round the answer to the nearest mile.

$$700 \div 1.06 \longrightarrow 70{,}000 \div 106$$

$$1.06 \times s = 700$$

$$\frac{\cancel{1.06} \times s}{\cancel{1.06}} = \frac{700}{1.06}$$

$$s = \frac{700}{1.06}$$

$$s = 660.3 \text{ miles per hour}$$

```
              660.3
      106 ) 70,000.0
           -63 6
             6 40
           - 6 36
              40
            -  00
              40 0
            - 31 8
               8 2
```

The speed of sound at an altitude of 42,000 feet is about 660 miles per hour.

Around the World

For hundreds of years, explorers searched for a route that would take them around the world. Before airplanes, only land and water passages were available. In the early 1900s, many pilots and airplane inventors raced to be the first to **circumnavigate** the globe in an airplane. The first person to fly around the world was Wiley Post in 1931 with the aid of **navigator** Harold Gatty. In 1933, Post became the first person to fly solo around the world. From July 15 to July 22, 1933, Post flew his Lockheed Vega monoplane *Winnie Mae* 15,596 miles (25,094 km) in 7 days, 18 hours, and 49 minutes. Post started and ended his flight at Floyd Bennett Field in Brooklyn, New York.

From February 26 to March 2, 1949, a crew of 14 made the first nonstop flight around the world. The crew, led by Captain James Gallagher, flew in a U.S. Air Force Boeing B-50 bomber named *Lucky Lady II*. During this famous flight, the B-50 bomber went through 4 in-flight refuelings! The *Lucky Lady II* made it around the world in 3 days, 22 hours, and 1 minute. This amazing test proved that the U.S. Air Force was capable of reaching any location on Earth within days.

From March 19 to April 17, 1964, female pilot Jerrie Mock became the first woman to make a solo flight around the world. Mock flew 22,858 miles (36,779 km) and made 21 stops.

Wiley Post

Lucky Lady II

Jerrie Mock

In 1986, pilots Dick Rutan and Jeana Yeager (no relation to Charles Yeager) were the first people to fly around the world without refueling. The plane that Rutan and Yeager flew—named *Voyager*—was designed by Rutan's brother, Burt. Burt Rutan created a plane that was lightweight, durable, and could carry a large amount of fuel. With all 17 fuel tanks full, the plane weighed about 9,700 pounds (4,404 kg)! Much of the fuel was stored in the wings.

Rutan and Yeager began their famous flight from Edwards Air Force Base in California at 8:01 a.m. on December 14, 1986. *Voyager* was so heavy with fuel that it almost did not make it off the ground. Flying was difficult and dangerous. The **cockpit** of the plane was about the same size as a small closet. Each pilot took a turn flying the plane while the other rested. Several storms threatened to destroy their plane, and they used up much-needed fuel steering around them. *Voyager* landed back at Edwards Air Force Base with only a few gallons of fuel left in its tanks at 8:06 a.m. on December 23, 1986. Not only were Rutan and Yeager the first to circumnavigate the globe in a plane without refueling, they also set a record for the longest nonstop flight: just over 216 hours.

Let's say that each of *Voyager*'s 17 fuel tanks were the same size, and each could carry 515.3 pounds of fuel. How much did *Voyager* weigh when its fuel tanks were empty?

List the information first:

Voyager's weight with fuel = 9,700 pounds
number of fuel tanks = 17
fuel each tank could carry = 515.3 pounds
Voyager's weight without fuel = v

The total weight of the plane is equal to the weight of the plane plus the weight of the fuel. To get the total weight of the fuel, we can multiply the number of tanks by the weight of fuel each tank could carry. Then, solve for v, the weight of the plane without fuel.

(17 x 515.3) + v = 9,700 ←

Parentheses tell us what needs to be done first.

```
      515.3
  x      1 7
   3 607 1
 + 5 153
   8,760.1
```

8,760.1 + v = 9,700
8,760.1 − 8,760.1 + v = 9,700 − 8,760.1
v = 939.9 pounds

Voyager weighed about 939.9 pounds when its fuel tanks were empty.

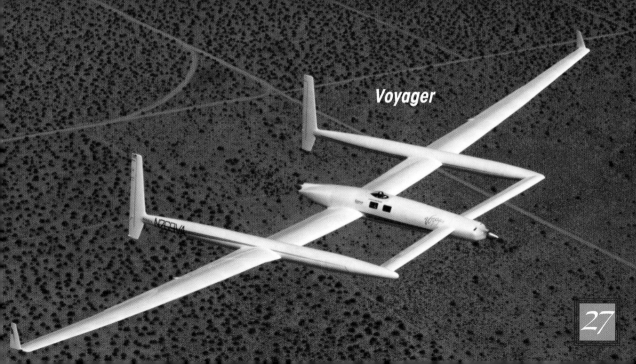

Voyager

Airplanes are not the only aircraft to have changed over the years. Hot-air balloons have come a long way since the first flight in 1783. Inventors have experimented with a wide variety of materials and designs.

In 1999, Bertrand Piccard of Switzerland and Brian Jones of England became the first people to circumnavigate the world in a hot-air balloon. The balloon—named the *Breitling Orbiter 3*—was made of the best materials modern science could offer. The balloon itself was 180 feet (55 m) tall when filled with air. Two methods kept the balloon in the air. An envelope at the top of the balloon contained **helium** gas. The rest of the balloon was filled with hot air created from burning **propane**. The journey began on March 1 and ended 19 days, 21 hours, and 55 minutes later.

Dirigibles, large airships that can be steered, became a popular means of transportation in the early 1900s, before the widespread use of airplanes. During World War I, some countries used dirigibles for military and transportation purposes. They can often hold many people and can stay in the air for extended periods of time. Several disasters involving airships resulted in decreased interest in this means of transportation. Today, some dirigibles called blimps are used for advertising and races.

The *Breitling Orbiter 3* traveled about 59.48 miles (95.7 km) per hour. What was the approximate total distance of the flight? Round to the nearest mile.

First, figure out the total number of hours traveled, rounding to the nearest hour. We know that the flight time for the *Breitling Orbiter 3* was 19 days, 21 hours, and 55 minutes. Multiply the number of days by the number of hours in a day.

19 x 24 = 456 hours ⬅

$$\begin{array}{r} 19 \\ \times\ 24 \\ \hline 76 \\ +\ 38 \\ \hline 456 \end{array}$$

Now, add the remaining time of the flight to 456 hours. We can round 21 hours and 55 minutes to 22 hours.

456 + 22 = 478 hours

Let's review our information:

approximate length of flight = 478 hours
speed of travel = 59.48 miles per hour
total distance = *d*

To find the total number of miles traveled, we can set up an equation based on the distance formula: distance = speed x time. Solve for *d*.

$d = s \times t$
$d = 59.48 \times 478$ ⬅

$$\begin{array}{r} 59.48 \\ \times\ \ \ 4\,78 \\ \hline 475\,84 \\ 4\,163\,6 \\ +\ 23\,792 \\ \hline 28{,}431.44 \end{array}$$

$d = 28{,}431.44$ miles

The *Breitling Orbiter 3* traveled about 28,431 miles.

The Future of Flight

In just over 100 years, brave inventors and pilots have broken nearly every barrier in aviation. Pilots have flown all over the globe, including flights to the North and South Poles. Today, the fastest jet plane is the SR-71 Blackbird, which can fly at speeds around 3.3 times faster than the speed of sound. NASA (National Aeronautics and Space Administration) has built an aircraft named the X-43 Hyper-X that may fly even faster. Commercial airliners carry thousands of people all over the world daily. The Airbus A380—the newest and largest commercial airliner—weighs about 300 tons (272 t) and can carry 555 passengers. Scientists have created spacecraft capable of traveling into space and then landing on Earth again the same way airplanes do. In a short time, aviation has come a very long way.

With all of these accomplishments, what is next for the world of aviation? Technology continues to improve and inventors continue to come up with new ideas and methods. Computers give pilots more control over planes than ever before. Scientists have already developed planes that can be flown by remote control from the ground. In the future, planes may not even need pilots anymore! Whatever may happen in the field of aviation, people will continue to use algebra and variables to study the famous flights of the past and future.

Glossary

aeronautical (ehr-uh-NAW-tih-kuhl) Pertaining to the science of operating aircraft.

altitude (AL-tuh-tood) The elevation of an object above the ground.

aviation (ay-vee-AY-shun) A field of science that deals with the design, manufacture, and operation of airplanes.

biplane (BY-playn) An airplane with 2 main sets of wings, usually with 1 positioned above the other.

circumnavigate (suhr-kuhm-NA-vuh-gayt) To go completely around something, such as Earth.

cockpit (KAHK-piht) The section of an airplane where the pilot works.

dirigible (DIHR-uh-juh-buhl) An airship that can be steered.

distinguished (dih-STING-wished) Marked by excellence.

helium (HEE-lee-uhm) A gas that is often used for balloons and is lighter than air.

monoplane (MAH-nuh-playn) An airplane with 1 main set of wings.

navigator (NA-vuh-gay-tuhr) A member of an airplane's crew responsible for plotting the course of the flight and keeping the plane on that course.

propane (PROH-payn) A gas that is easily burned and is often used as fuel.

technology (tek-NAH-luh-jee) Processes, methods, and knowledge for accomplishing a task.

transatlantic (trans-uht-LAN-tik) Relating to or involving crossing the Atlantic Ocean.

Index